I0480291

A New Approach for Understanding the Mechanism and Drug Design of Chronic Myeloid Leukemia Responsive Protein

An In-Silico Study for the Design of Isoform Specific Anti-Leukemic Agent

Hridoy Ranjan Bairagya
Gyan Prakash Rai
Alvea Tasneem
Saima Reyaz

ELIVA PRESS

ELIVA PRESS

Hridoy Ranjan Bairagya
Gyan Prakash Rai
Alvea Tasneem
Saima Reyaz

Chronic myeloid leukemia (CML) is a blood cancer. In CML cancer, stem cells of bone marrow form abnormal WBCs that hinder the function of normal RBCs and WBCs. CML occurs due to a chromosomal translocation between BCR gene on Chromosome-22 and ABL gene on Chromosome-9. The purine nucleotide biosynthesis pathway consists of three enzymes (hIMPDH, hGMPS and hGMPR) which are identified to be responsible for CML cancer and they are also involved in cellular metabolic pathways that exhibit elevated levels of activity in rapidly proliferating cells, such as neoplastic and regenerating tissues. These CML responsive proteins exist as type-I and II isoforms. Isoform-II is responsible for CML cancer, whereas isoform-I which keeps a housekeeping role, and observed in normal cells. Our in-silico methods may explore some new biochemical mechanism and novel conformation of this protein that may be effective for future drug discovery. The basic problem for discovery of drugs for CML cancer protein is that all proposed inhibitors bind to CML cancerous and normal protein, as a result both normal and cancer cells lose their function and die. To design the drugs for CML cancerous protein, isoform specific (type-II) drug design techniques were employed in this study and the new inhibitors were identified by computational method using different databases. The computationally designed proposed drugs can be effective to recognize the CML cancer protein and may act as a good drug candidate for CML cancer.

Published: Eliva Press SRL
Address: MD-2060, bd.Cuza-Voda, 1/4, of. 21 Chişinău, Republica
Moldova
Email: info@elivapress.com
Website: www.elivapress.com

ISBN: 978-1-952751-51-6

© Eliva Press SRL
© Hridoy Ranjan Bairagya, Gyan Prakash Rai, Alvea Tasneem,
Saima Reyaz
Cover Design: Eliva Press SRL

No part of this book may be reproduced or utilized in any form or by
any means, electronic or mechanical, including photocopying,
recording, or by any information storage and retrieval system,
without permission in writing from Eliva Press.

All rights reserved.

A NEW APPROACH FOR UNDERSTANDING THE MECHANISM AND DRUG DESIGN OF CHRONIC MYELOID LEUKEMIA RESPONSIVE PROTEIN

An In-Silico Study for the Design of Isoform Specific Anti-Leukemic Agent

Hridoy R. Bairagya[1*], Gyan Prakash Rai[2], Alvea Tasneem[2] and Saima Reyaz[2]

[1]Department of Biophysics, All India Institute of Medical Sciences,

New Delhi-110029, India

[2]Department of Computer Science, Jamia Millia Islamia,

New Delhi-110025, India

*Corresponding Author:
Dr. Hridoy Ranjan Bairagya
(Department of Biophysics, All India Institute of Medical Sciences, New Delhi-110029)
Tel: +91-8277145815
E-mail: hbairagya@gmail.com

ABSTRACT

Human Guanosine-5'-Monophosphate Reductase (hGMPR) enzyme is involved in purine nucleotide biosynthesis pathway and its isoform-II promotes monocytic differentiation of HL-60 leukemia cells. The inhibition of hGMPR may lead to the apoptosis of neoplastic cell lines and induce differentiation in chronic myeloid leukemia (CML) cells after blast transformation. So, hGMPR-II attracts an excellent potential target for design of isoform specific anti-leukemic agents. For the purpose of designing of anti-cancer agent, computational based analyses were performed on the several crystal structures of hGMPR enzyme. Molecular Dynamic (MD) simulation provides the new insights about the catalytic mechanism of enzyme, which lead to the involvement of four conserved water molecules at substrate (GMP) binding domain in native or unliganded free conformation. In X-ray structures, the interaction of active site residue Cys186 with Thr188/Glu289 occurs as Cys186---GMP---Thr188/Glu289 in ligand bound state. However, MD simulation studies suggest that above mentioned catalytic residues move away from the substrate GMP binding pocket and they are stabilized by four conserved water molecules as Cys186---W1---W2---W3---W4---Thr188/Glu289. Therefore, these four conserved water molecules may stabilize that GMP binding pocket in native conformation, but they move away and vacate their positions when enzyme (hGMPR)-substrate (GMP) complex is formed. Thus, our in-silico studies highlight the structural importance of conserved water molecules which may play a curtail role for inhibitor binding mechanism at GMP binding site. For designing the isoform specific inhibitor for hGMPR-II, virtual screening technique and molecular docking study has been applied. The computational investigations especially pharmacological, toxicological and computational docking study suggest compound ZINC_2522481 and ZINC_4963099 have a better drug score compared to substrate GMP. Both these compounds bind tightly at substrate bind site in cancerous protein than normal one, and they may act as a better drug candidate for CML cancer protein.

1. INTRODUCTION

Chronic Myeloid leukemia (CML) is the most common blood cancer worldwide with incidence rate estimated to be 1 to 2 cases out of 100,000 populations per year. It occurs to all age groups but is predominantly a disease of older people, and remains uncommon in younger individuals [1]. In India, annual incidence of CML is reported to be 0.8 to 2.2 per 100,000 population with age adjusted rate (AAR; per 100,000) of 0.71 in males and 0.53 in females [2].

CML is caused due to chromosomal translocations, in which ABL1 (Abelson murine leukemia) gene on chromosome 9 and the BCR (breakpoint cluster region) gene on chromosome 22 fused together which then generates Philadelphia (Ph) chromosome. This fusion result in expression of an oncogenic protein named as BCR-ABL1. BCR-ABL1 is an active fusion tyrosine kinase that regulate by the downstream signaling pathways such as RAS, MAPK (**M**itogen-**a**ctivated **p**rotein **k**inase), PI3 (**P**hosphoinositide-3) kinase, MYC (**M**yelocytomatosis oncogene cellular homologue) and JAK-STAT (Janus kinases; **S**ignal **t**ransducer and **a**ctivator of **t**ranscription) [3-7], which encourage growth and replication. This result in leukemogenesis by creating a cytokine independent cell cycle with abnormal apoptotic signals in response to cytokine withdrawal [8]. Some symptoms of CML are hepatomegaly, fatigue, weakness, dragging pain, pallor, or sometime asymptomatic are observed.

For the treatment of CML cancer, various anti-neoplastic agents, such as, imatinib, nilotinib and dasatinib; some other agents like hydroxyurea, interferon alpha, busulfan, etc.; were led to profound benefit in survival and quality of life for both adult and young patients. Through the history, the disease is characterized by progression through three phases, chronic phase, accelerated phase and blast crisis. The patients at chronic phase show asymptomatic symptoms like fatigue, early satiety, or complications of hyperviscosity such as visual disturbances or priapism. The chronic phase is characterized by proliferation of WBCs (white blood cells) and sometimes platelets and splenomegaly. Symptoms can be controlled by agents such as hydroxyurea, interferon alpha, or imatinib. However, these agents cannot prevent the progression to accelerated phase, where progressive loss of WBCs differentiation with increase in the number of immature blast cells, which then eventually leads to blast crisis occurs, resulting in imminent death [WHO, 2014] [9]. Possibly, the scientific reason behind this mystery of unavailability of drugs for CML cancer is that they failed to recognize the isoform-specific proteins.

Understanding the distinction of identical twins are a difficult and peculiar tasks, while in the massive biological system differentiation of isozymes remains a main challenge in biochemical science, unless its dynamic behavior predict the proper conformations in long time computer simulation processes. The structure, function, domain recognition and kinetic properties of both isozymes in human GMPRs are very resembling, so that one could not predict the actual functional morphology of any particular system. Therefore inhibition of hGMPR-II beyond stirring the isoform-I, is the fundamental intricacy in the Nature. In the guanine nucleotide pathway, Inosine Monophosphate Dehydrogenase (IMPDH, EC 1.1.1.2051) catalyzes the oxidation of IMP (Inosine Monophosphate) to produce XMP (Xanthosine Monophosphate), while Guanosine Monophosphate Synthetase (GMPS, EC 6.3.5.2) converts the XMP to GMP (Guanosine Monophosphate) and Guanosine Monophosphate Reductase (GMPR, EC 1.7.1.7) deaminate the GMP into IMP (Figure 1) [10]. Consequently, all these three enzymes (hIMPDH, hGMPS and hGMPR) are involved in cellular metabolise pathways that exhibit elevated levels of activity in rapidly proliferating cells, such as neoplastic and regenerating tissues. The human GMP

5

reductase enzyme exists in two isoforms: isoform-I is expressed in normal cells and has a housekeeping role whereas isoform-II is expressed in leukemic cancer cells (because over expression promotes monocytic differentiation of HL-60 leukemia cells) [11]. Hence isoform-II is an excellent potential target for design and development of isoform specific anti-leukemic agent.

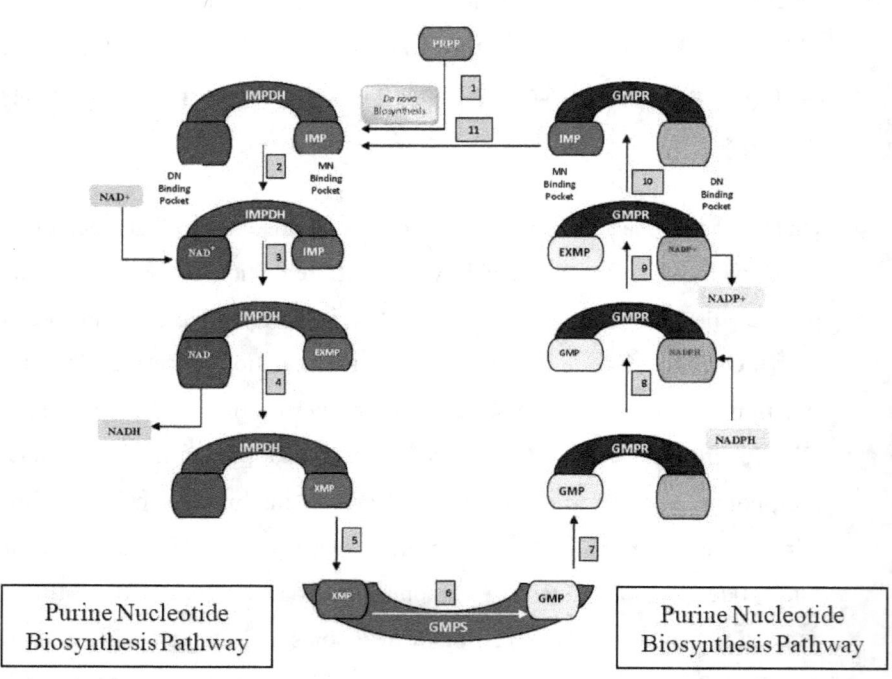

Figure 1 : Involvement of the hIMPDH, hGMPS and hGMPR enzymes in the purine nucleotide biosynthesis pathway for interconversion of IMP to XMP to GMP.

IMP: Inosine Monophosphate, XMP: Xanthosine Monophosphate, and GMP: Guanosine Monophosphate.

The catalytic domain of hGMPR is folded to form two specific nucleotide binding motifs. One of them accommodates substrate or mononucleotide or GMP, and another contains cofactor, di-nucleotide or NADPH (Nicotaminde Adenine Dinucleotide Phosphate) (Figure 2).

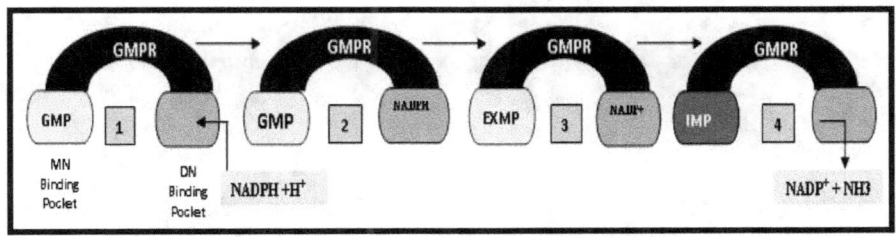

Figure 2: Conformational changes and transition states of human GMPR enzymes

The hGMPR is a homotetramer with 37-kDa subunits and monomer of this enzyme has a $(\alpha/\beta)^8$ structure of TIM barrel (Figure 3 (A)) [12]. The hGMPR enzyme maintains intracellular balance between the adenine and guanine nucleotide pools and it also participates in re-utilization of free intracellular bases and purine nucleosides. Few crystal structures (PDB Id: 2BLE, 2C6Q and 2BZN) [13] of hGMPRs enzyme are available in different conformations, but it may not address the structural changes of the enzyme. The proposed biochemical mechanism of substrate to product of hGMPR enzyme catalyzes two different chemical transformations- first is deamination whereas second one is hydride ion transfer process. In deamination, the thiol group (-SH) of catalytic Cys186 attacks C2 of substrate GMP to form a covalent thioimidate intermediate (E-XMP). In the second step, nicotinamide (NADPH) moves adjacent to C2 of E-XMP to allow hydride ion transfer, resulting in production of IMP (Figure 3 (B)) [14].

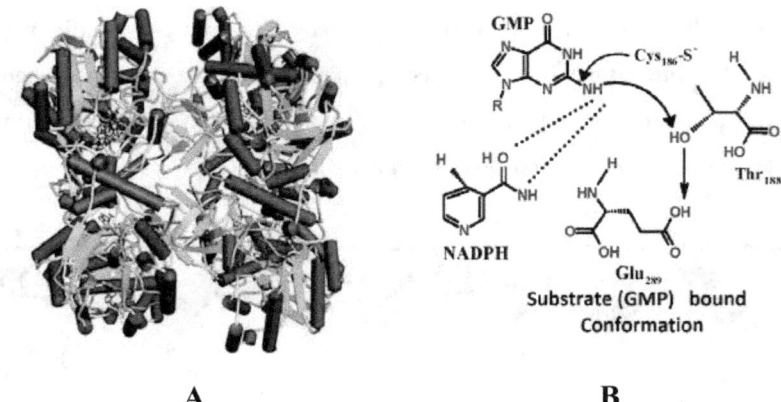

Figure 3: (A) X-ray structure (PDB Id: 2C6Q) of hGMPR in complex with IMP and NADPH. (B) Schematic representation of GMPR enzyme showing Cys186 attacking C2 of GMP to form E-XMP intermediate.

Conserved water molecules play some crucial roles in enzymes/proteins and it has been known for some time. The function of water molecules are condensation and hydrolysis reactions and act as a critical molecular element—a product or reactant. In biochemical reactions water molecules may also function as transition state intermediates and it acts at the molecular level as a structural element, interconnecting the protein through hydrogen bonds, salt bridge interactions and maintaining/stabilizing the positions of residues and/or the protein fold [15]. These multiple roles played by water molecules are associated with its unusual and unique properties; its small size, the dipolar nature caused by its charge distribution, the capacity to act both as a hydrogen bond donor and acceptor, and the entropic gain associated with the release to bulk solvent, when bound to proteins and ligands. These structural waters are often conserved within a same protein family, that is, they perform the same functions and occur in nearly identical three-dimensional locations with reference to their associated structures [16]. More specially, the nucleotide binding sites of the enzyme are stabilized by one conserved water molecule [17] and it may associate for electron transfer mechanism between them. Arg322 anchoring both these sites of the enzyme through a conserved water molecular triad, which is a very unique in nature [18]. Moreover, loop and flap regions are also controlled by three conserved water molecular centers [19]. Water mediated salt bridge interactions are also involved at inter- or intra-domain recognition and stabilization [20]. Interestingly, another nucleotide binding GAT domain of hGMPS enzyme contains oxyanion pocket that is occupied by negatively charged atom of ligand in ligand bound state, in fact four conserved water molecules inactivate that pocket in unliganded state. The conserved sites of water molecules near the oxyanion hole highlight the structural importance of water, and suggest changes in the conventional definition of the chemical geometry of inhibitor binding site, its shape and complimentary [10]. Structurally conserved waters have been found in several classes of proteins, including nucleotide binding protein. In these studies conserved waters are often found in or near an hGMPR's active site or substrate GMP binding region, suggesting they play an important function in active site stability, flexibility, ligand coordination and residue positioning, hence their evolutionary conservation.

In spite of structural distinction between two isozyme, any experimental or computational studies have not been carried out on hGMPR protein to identify the native conformation of hGMPR enzyme. Thus science has developed most powerful sophisticated biophysical tool called "Molecular Dynamics Simulation" that not only deals low energy landscape of complex biological system but also its inner visions guide to solution of selective inhibitor design. Today, use of computational processing power, accuracy and its prompt responses are blooming for human mankind in different prospects of life science. Dynamics is a powerful theoretical method routinely used to simulate the dynamics of physical, chemical and biochemical systems. A deal of modern theoretical research is based on results of such simulations and in development of new algorithms to extend range of such simulation methods to large systems

9

and to longer time [21]. Therefore, the approach of MD simulation study is the only option that may provide the important information about the structural and functional insights in native conformation of GMP binding domain and role of conserved water molecules and stabilization of that site. To accomplish this, we have performed molecular dynamics simulation to investigate the novel conformation of GMP binding pocket that may be only evolved during MD simulation.

Apart from identifying the conserved water molecules at substrate binding pocket, comparative molecular docking analysis was also carried out on ligand free X-ray structures of hGMPR-I (2BLE) and II (2C6Q) to investigate isoform specific inhibitors that are structural analogs of substrate GMP or product IMP. As we know, the isoform-II of hGMPR is expressed in cancerous cells, so this proposed selected ligands may inhibit the cancerous protein (isoform–II) rather than the normal protein (isoform-I), as a result reducing the activity of the cancer cell. In order to identify the isoform specific inhibitor for hGMPR-II, we have employed virtual screening to select the specific ligands from reputed scientific database and have performed molecular docking simulation on both the X-ray structures of isoform I and II because in-silico molecular docking is one of the most powerful and well known techniques that is used for structure-based drug design [20]. The pharmacological parameters for example QSAR, ADME and toxicology properties were also been calculated to achieve drug likeness score.

2.   MATERIALS AND METHODS

2.1   Starting structure

The atomic coordinates of three crystal structures of hGMPR (isoform-I: PDB Id. 2BLE, and isoform-II: PDB Id. 2C6Q and 2BZN) enzyme were obtained from Protein Data Bank [22]. These PDB structures have been solved in different space groups with different resolutions. The "A" molecules from 2BLE (hGMPR-I), 2C6Q and 2BZN (hGMPR-II) were obtained from their respective X-ray structures using the Swiss PDB viewer program [23]. To discriminate structural, functional and dynamical characteristics of the native conformation of two isoforms, 2BLE, 2C6Q and 2BZN were taken as templates because their structural qualities and resolutions are better than the remaining crystal structures. Furthermore, the catalytic (Cys186) and non catalytic (Thr188 and Glu289) residues and its surrounding interaction of all X-ray structures were also analyzed using Swiss PDB Viewer program.

2.2   Sequence Similarity and Identification of residues of substrate and product binding pocket in X-ray structures

The protein sequences of hGMPR-I (PDB Id: 2BLE) and II (PDB Id: 2C6Q) were obtained from the UniProt database [24] and they were aligned using Clustal Omega (v1.2.4) program [25]. In addition, H-bonding interaction between substrate GMP and catalytic residues of Cys186, Thr188 and Glu289 and non catalytic residues Ser184, Asp219, Gly221, Met240, Gly242, Gly243, Met269, Ser270, Arg286 and Ser288 in 2BLE have been measured accordingly. Similarly, interactions between IMP and non catalytic residues of Ser184, Asp219, Gly221, Met240, Gly242, Gly243, Met269 and Ser270 have also been measured in 2C6Q X-ray structures (shown in Figure 4).

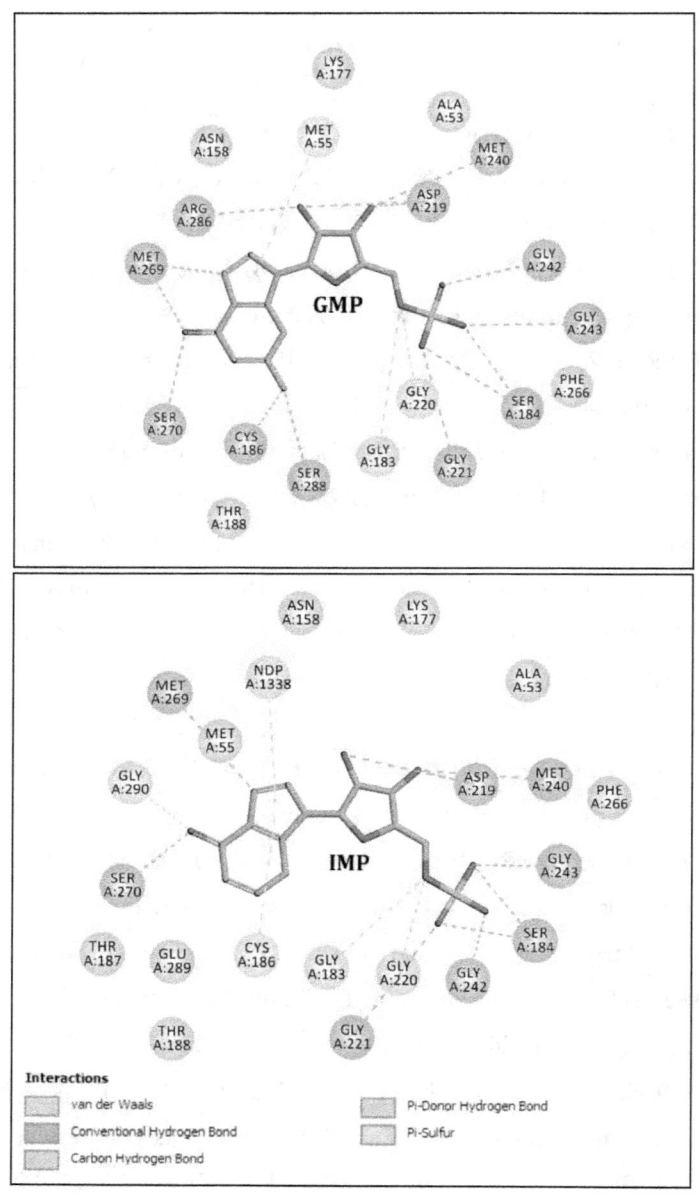

Figure 4: A 2-dimensional structure of amino acid residues interactions with surrounding ligands GMP and IMP in hGMPR-I (top) and II (bottom) respectively.

2.3 Adding missing residues and creating unliganded conformation

The "A" molecules of protein (excluding crystal water molecules) were obtained from the respective crystal structures of hGMPR (PDB Id: 2BLE, 2BZN and 2C6Q) for molecular dynamics simulation. The missing residues of amino acids sequences were added at 2BLE (257-260, 279-289 and 344-345), 2BZN (1-8, 279-286 and 337-341) and 2C6Q (1-8 and 337-341) corresponding crystal structures. Each residues were added to the appropriate positions at the corresponding structures and energy minimization was performed using 500 steps of steepest decent [26] followed by 1000 cycles of conjugate gradient [27] without constraints using Swiss PDB Viewer program. To create a native conformation of human GMPR, the crystal water molecules and ligands were remove from their respective crystal structures. Then the entire model structure was re-energy minimized (500 steps of steepest descent followed by 1000 cycles of conjugate gradient) until the energy charges decreased to 0.001 kJ/mol, for a 10 Å cutoff distance [28] in the non-bonded interaction, with a distance-dependent dielectric constant.

2.4 Molecular Dynamics simulation of hGMPR

The ligand free model structures of 2BLE (hGMPR-I) , 2BZN and 2C6Q (hGMPR-II) were considered for MD simulation excluding crystal water molecules and then missing hydrogen atoms were added to structure using the AutoPSF module of VMD (v1.9.3) program [29]. This structure was then solvated in a cubic box of TIP3P water molecules [30] extending at least 10 Å from protein surface [31] then sodium and chloride ions were used to neutralize the overall charge of the system. Water molecules and ions of the solvated system was initially minimized for 1000 cycles using CHARMM36 force field [32] and then whole system has been energy minimized by 1000 cycles of steepest descent method. Then the system has been gradually heated from 273K to 310K over 100ps under *NVT* conditions. Then system has been finally equilibrated in *NPT* conditions for 500 ps with an integration time step of 2 fs [33]. The final production run has been carried out for all atom simulation in the *NPT* ensemble at 310K for a time period of 30ns at a time step of 1 fs using NAMD (v.2.9) program [34]. Periodic boundary conditions and a cutoff distance of 12Å, switch distance of 10Å, pairlist distance of 14Å for van der Waals interactions were applied. The atomic coordinates of MD structures were recorded at every 2ps for trajectories analysis.

2.5 Calculation of residential frequency and coordination of conserved water centers

The 25,000 recorded snapshots (50ns) of MD structure were critically analyzed to investigate the interactions of catalytic residues (Cys186, Thr188 and Glu289) with four conserved water molecules (W1, W2, W3 and W4), those were occupying at the substrate binding site. These four conserved water molecules (W1 to W4) were identified from different native MD structures of the substrate binding domain using Swiss PDB Viewer program; in addition, the conserved positions of these water molecules were also verified by the 3dSS program [35]. In between two water molecules whose centre-to-centre (oxygen atoms) distance (in different snapshots) was within 1.8 Å [36] in between reference and movable structure was assigned as conserved. The initial MD frame from each structure was taken as a reference and other prerecorded snapshots (frames from 50 to 100ns) were assigned as movable and superimposed on that initial frame to identify the conserved water molecules between initial and movable fame. The W1 of Cys186, W2 of W1 and W3, W3 of W2 and W4 and W4 of Thr188/Glu289 dyad were assigned as conserved water positions. Hence, these four conserved water molecules were assigned new numbering scheme from W1 to W4 for maintaining the clarity from water molecules of crystal structures.

2.6 Virtual Screening

The isomeric SMILES of ligand GMP and IMP were obtained from the Drug Bank database [37] and they were used for high throughput screening (HTS) techniques to investigate structural analogs of GMP or IMP by Swiss Similarity program [38]. Five different chemical libraries or databases (FDA approved [37], ligands from PDB [39], ChEMBL [40], zinc-drug like and zinc-lead like [41]) were selected for HTS method to achieve best structural analogs of GMP and IMP. After screening from aforesaid chemical libraries, fifty structural analogs of substrate GMP or product IMP have been isolated with reasonably good similarity score of GMP/IMP and finally fifteen molecules or compounds were considered. The molecular id of GMP or IMP structural analogs are DB01972, DB03315, DB03593, DB04457, ChEMBL752, ChEMBL460901, ChEMBL582887, GAO, 1MG, OMG, ZINC_2522481, ZINC_4963099, ZINC_13531931, ZINC_3869846 and ZINC_4228242 (as mentioned in Table 1).

2.7 Pharmacokinetics properties and QSAR predictions

A pharmacokinetics property involves drug-likeness, which is a prediction to determine whether a particular compound has some properties consistent with being orally active drug [42]. This concept was established by Lipinski et al [43], called Lipinski rule of five. Another property involves the drug score, which is a parameter that determines whether the compounds can be selected as a drug candidate or not [44]. The higher drug score value means higher chance as a drug candidate [45]. The Osiris property explorer [46] and Swiss-ADME predictors [47] programs were used for this prediction of in-silico drug likeness and toxicity predictions of the screened ligands. Each molecule was filtered out on the basis of six molecular properties (cLogP, solubility, molweight, TPSA, drug likeness and drug-score) from Osiris program and four characters (molar

15

refractivity, Lipinski, bioavailability score and synthetic accessibility) from Swiss ADME program. Analyses of the compounds were compared with that of substrate GMP or product IMP, and those selected compounds without violation of pharmacokinetic properties were used for further molecular docking study.

2.8   Structure Preparation and Molecular Docking

Molecular docking is a methodology for calculating the binding interactions between bio- molecules like protein or enzyme (receptor) and small molecules (ligands) to form a stable complex for drug design. Docking is a method for fitting of ligand into an appropriate binding site of a target receptor with non-covalent bonding in order to formed stable compound. The ligand-protein interactions are determined by the term binding free energy through the use of various scoring functions [48] such as force-field based [49], empirical [50], or knowledge-based scoring functions [51].

Here, the native conformation X-ray structure of 2BLE (open loop conformation) and 2C6Q were taken (excluding water molecules) as a receptor for molecular docking studies using the AutoDock Vina program (v 1.5.6) [52]. Before performing the docking studies, the X-ray structure of 2BLE is occupied predominately by GMP, but the cofactor binding site is empty because that location is already occupied by a loop (His/Tyr278 to Gly290) region. The loop may hinder the docking of the ligand or compound, hence, the conformation of this loop region was reoriented similar to 2C6Q to open the surface of that pocket. Required PDBQT file of each protein and ligand molecule were assigned Kollman united atom charges and Gasteiger partial atomic charges respectively [53]. All the rotatable bonds of each ligand were kept to be flexible during docking simulation, and residues inside the binding pockets make rigid [54]. Grid point spacing was set at 1 Å and grid points 22 were taken for 2BLE and grid points 23 were taken for 2C6Q in each direction of grid box respectively and both the boxes were centered at -SG of Cys145. The remaining docking parameters were assigned to their default values. We obtained the best five docking results (from each docked complexes) according to their binding energy of protein. The best and appropriate conformation (with respect to binding free energy) of each docked compound was selected by monitoring their interaction with 2BLE and 2C6Q protein as similar with their corresponding crystal structures. Furthermore, a comparative analysis was carried out for finding the best structural analogs that can strongly bind at the substrate binding site of the cancerous protein (2C6Q) whereas it will loosely bind at the normal protein (2BLE).

Library	Sl.no.	Ligand Id	Name	Structure
FDA Experimental Drug	1	DB01972	Guanosine-5'-Monophosphate	
	2	DB03315	Guanosine-3'-Monophosphate	
	3	DB03593	N7-Methyl-Guanosine-5'-Monophosphate	
	4	DB04457	2'-Deoxyguanosine-5'-Monophosphate	
ChEMBL	5	CHEMBL752	Adenosine Phosphate	
	6	CHEMBL460901	$C_9H_{13}N_2O_9P$	
	7	CHEMBL582887	$C_9H_{14}N_3O_8P$	

Ligands from PDB	8	GAO	Guanine Arabinose-5'-Phosphate	
	9	1MG	1N-Methylguanosine-5'-Monophosphate	
	10	OMG	O2'-Methylguanosine-5'-Monophosphate	
Zinc Drug-Like	11	ZINC_2522481	2-amino-9-[(2S,3S,4R,5S)-3-hydroxy-4-methoxy-5-methylol-tetrahydrofuran-2-yl]hypoxanthine	
	12	ZINC_4963099	2-amino-9-[3-hydroxy-5-(hydroxymethyl)-4-methoxy-tetrahydrofuran-2-yl]-3H-purin-6-one	
Zinc Lead-Like	13	ZINC_13531931	Ara-HxMP	
	14	ZINC_3869846	2'-Deoxyguanosine-5'-monophosphoric acid disodium salt	
	15	ZINC_4228242	Disodium 5'-Inosinate	

Table 1: List of structural analogs of GMP/IMP from Swiss Similarity Program

3. RESULTS AND DISCUSSION

3.1. Sequence Similarity between hGMPR-I and II

The sequence of hGMPR-I and II isoforms have 90% similarity. The residues of Lys9, Val52, Ser68, His70, Val101, Gln107, Asn108, Ser115, Thr190, Cys251 and Ala252 of hGMPR-I was observed to be replace by Ser9, Ala52, Cys68, Phe70, Ala101, Ser107, Ser108, Gln115, Lys190, Ser251 and Gly252 of isoform-II. The mismatch residues have marked in red colour in Figure 5. The substrate binding pocket is enclaved by Ser184, Cys186, Asp219, Gly221, Gly242, Gly243, Met269, Ser270, Arg286, Ser288 and Gly290 residues, those are observed in both isoforms, as a result no mismatch key residue are found. However, one key residue (resid 105) at cofactor binding site is located. The Ser105 of hGMPR-I is replaced by Thr105 in hGMPR-II, the structural and geometrical features of this residue may be used for isoform based drug design.

```
hGMPR-I    MPRIDADLKLDFKDVLLRPKRSSLKSRAEVDLERTFTFRNSKQTYSGIPIIVANMDTVGT    60
hGMPR-II   SSGLVPRGSLDFKDVLLRPKRSTLKSRSEVDLTRSFSFRNSKQTYSGVPIIAANMDTVGT    60
             :   ***************:****:**** *:*:**********:***.********

hGMPR-I    FEMAAVMSQHSMFTAIHKHYSLDDWKLFATNHPECLQNVAVSSGSGQNDLEKMTSILEAV    120
hGMPR-II   FEMAKVLCKFSLFTAVHKHYSLVQWQEFAGQNFDCLEHLAASSGTGSSDFEQLEQILEAI    120
           ****  *:  :.*:***:****** :*: ** ::*:**:::*.***:*.*:*:: .****:

hGMPR-I    PQVKFICLDVANGYSEHFVEFVKLVRAKFPEHTIMAGNVVTGEMVEELILSGADIIKVGV    180
hGMPR-II   PQVKYICLDVANGYSEHFVEFVKDVRKRFFQHTIMAGNVVTGEMVEELILSGADIIKVGI    180
           ****:****************** **  :**:***************************:

hGMPR-I    GPGSVCTTRTKTGVGYPQLSAVIECADSAHGLKGHIISDGGCTCPGDVAKAFGAGADFVM    240
hGMPR-II   GPGSVCTTRRKTGVGYPQLSAVMECADAAHGLKGHIISDGGCSCPGDVAKAFGAGADFVM    240
           *********:************:****:**************:*****************

hGMPR-I    LGGMFSGHTECAGEVIERNGRKLKLFYGMSSDTAMNKHAGGVAEYRASEGKTVEVPYKGD    300
hGMPR-II   LGGMLAGHSESGGELIERDGKKYKLFYGMSSEMAMKKYAGGVAEYRASEGKTVEVPFKGD    300
           ****::**:*.. **:***:*:* ******** :*.*:*:*****************:***

hGMPR-I    VENTILDILGGLRSTCTYVGAAKLKELSRRATFIRVTQQHN    341
hGMPR-II   VEHTIRDILGGIRSTCTYVGAAKLKELSRRTTFIRVTQQVN    341
           **:** *****:******************:********* *
```

Figure 5: Sequence alignment between the two isoforms hGMPR-I and II. The key residues of substrate (S184, C186, D219, G221, G242, G243, M269, S270, R286, S288, G290)/ product (S184, C186, D219, G221, G242, G243, M269, S270, E289) and the cofactor binding pocket (K78, T105, S270, Y285, R286) are marked in pink and blue respectively. The identical /similar, strongly dissimilar, and non-matching residues are shown in green, white and red color

3.2. Molecular Dynamics of hGMPR

The Average root mean square deviations (RMSD) of each native MD structure (25,000 frames) were calculated using VMD program. The RMSD (CA atom) from final 50ns (50 to 100ns) of each MD simulation structure of hGMPR I and II were mentioned in Figure 6. However, their comparative trajectory analysis of RMSD (CA atom) between MD structures of 2BLE and 2C6Q indicates that the RMSD is within 0.58 to 2.59Å in 2BLE and 1.25 to 3.54 Å in 2C6Q. Interestingly, the superimposed graphical presentation of average RMSD of 2BLE and 2C6Q indicates the RMSD value is ~1.5Å in 2BLE and 3.5 Å in 2C6Q during the time period from 25 to 50ns. Hence, the backbone fluctuations are higher in 2C6Q than 2BLE that explores cancer protein (2C6Q) is more flexible and dynamic than normal protein (2BLE).

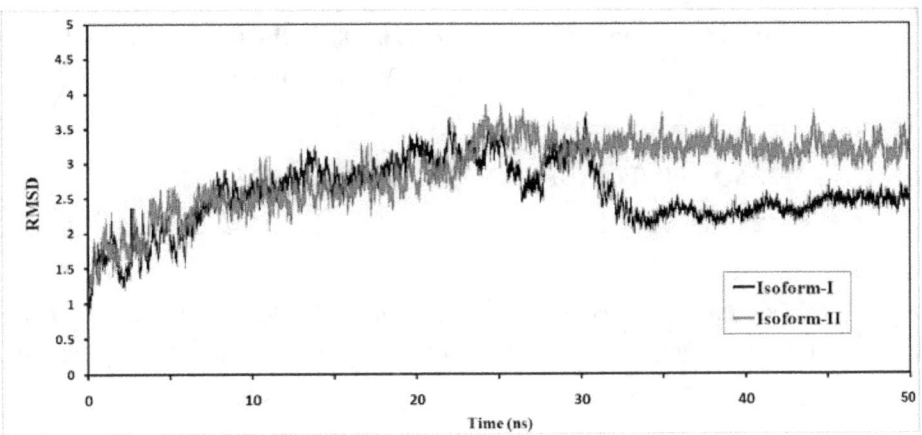

Figure 6: Presentation of RMSD (CA atom) of isoform-I (PDB Id: 2BLE) (black colour) and isoform-II (PDB Id: 2C6Q) (red colour) in native MD structures

21

3.3. Comparative analysis of substrate binding pocket with catalytic residues in crystal structures

The crystal structures of human GMPR enzyme in the protein data bank involve either substrate or product bound conformations. Our comparative analysis on three different crystal structures (2BLE, 2BZN and 2C6Q) of the protein revealed that C2 atom of substrate GMP lies at a distance of 4.15 Å from Cys186 (SG), 7.13 Å from Thr188 (OG) and 10.31 Å from Glu289 (OE) in the crystal structure of 2BLE. After the product formation, in the IMP binding site, the distance between C2 atoms of IMP show H-bond distances (< 3.50 Å) with Cys186 and Thr188. However the Glu289 (OE) was found to interact with NH_2 group of NADPH in the crystal structure of 2C6Q. Consequently, in 2BZN crystal structure, the distances from C2 atom of IMP to the catalytic residues mimics the similar results as in 2C6Q (Table 2).

PDB Id	Ligand	Cys186 (SG) (Å)	Thr188 (OG) (Å)	Glu289 (OE) (Å)	Thr188 (OG) to Glu289 (OE) (Å)
2BLE	GMP	4.15	7.13	10.37	13.10
2C6Q	IMP/NADP	2.31	3.88	5.97	2.27
2BZN	IMP	2.59	4.15	5.76	2.24

Table 2: Interaction between C2 atom of ligand (GMP/IMP) with side chains of catalytic residues (Cys186, Thr188 and Glu289). Distances are in angstrom (Å)

3.4. Identification of conserved water molecules in native conformation

To identify the conserved water molecules in hGMPR, the three different X-ray crystal structures (2BLE, 2BZN and 2C6Q) of protein were compared with their respective native conformations of MD structures. The crystal structure of 2BLE was taken as template and was superimposed (backbone atom) with native conformation MD structure of 2BLE. As can be seen from Figure 7, the catalytic residues Thr188 and Glu289 had departed from their initial position (in x-ray) and in their place a W3 and W4 water molecules are seen stabilizing their positions. Whereas the thiol (-SH) group of Cys186 has also changed its structural conformation and is in H-bonding distance with a W1 water molecule instead of C2 atom of GMP.

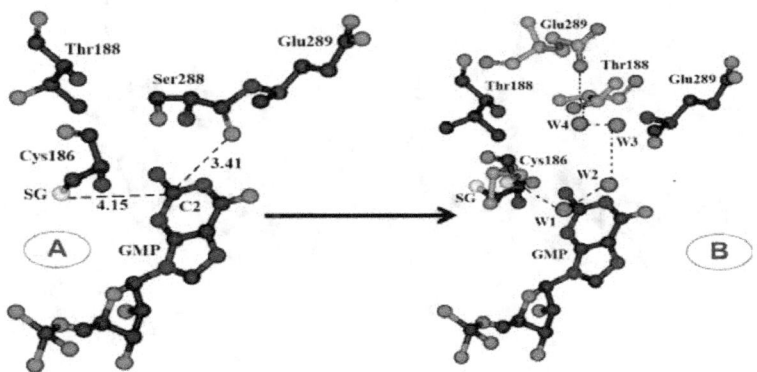

Figure 7: (A) The interaction of substrate (GMP) with catalytic residues in X-ray structure (PDB Id: 2BLE) of hGMPR enzyme. (B) Superimposed (backbone atoms) structure of hGMPR from X-ray (2BLE) and its native conformation structure from MD simulation are represented by green and CPK colors

For product (IMP) and cofactor bound conformation, the crystal structure of 2C6Q was taken as template and was superimposed (backbone atom) with native conformation MD structure of 2C6Q. As can be seen from Figure 8, the residues Cys186, Thr188 and Glu289 have departed from their initial position during MD simulation and Glu289 occupied the position near oxygen atom (O6) of IMP and –CONH$_2$ group of NADP. For product (IMP) bound conformation, the crystal structure of 2BZN was taken as template and was superimposed (backbone atom) with native conformation MD structure of 2BZN. As can be seen from Figure 9, during MD simulation, residues Thr188 and Glu289 has occupied the positions of Cys186 at IMP binding site whereas Cys186 also moved away from its initial position.

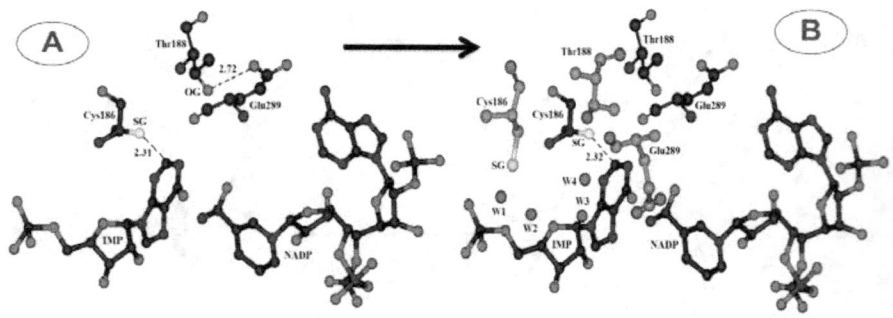

Figure 8: (A) In product (IMP) and cofactor bound conformation, the interaction of IMP with catalytic residues in X-ray structure (PDB Id: 2C6Q) of hGMPR enzyme. (B) Superimposed (backbone atoms) structure between X-ray of hGMPR (2C6Q) and native conformation of MD structure are represented by green and CPK colors

24

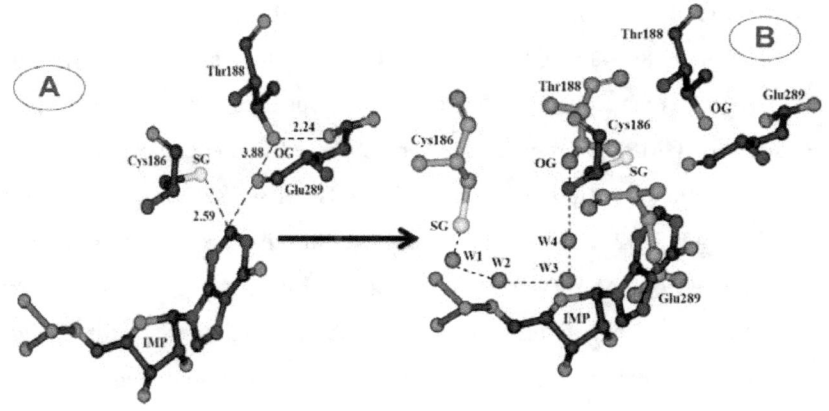

Figure 9: (A) The interaction of product (IMP) with catalytic residues in X-ray structure (PDB Id: 2BZN) of hGMPR enzyme. (B) Superimposed (backbone atoms) structure between X-ray of hGMPR (2BZN) and native conformation of MD structure are represented by green and CPK colors

3.5. Pharmacokinetics and Toxicity properties of structural analogs of GMP

Though we have identified the conserved water molecules at substrate GMP binding site in MD structure but especially no attempt has been made to search the structural analogs of GMP or IMP for isoform specific inhibitor design. Therefore, the virtual screening technique has been performed to identify the best structural analogs of GMP or IMP and further their pharmacokinetics and toxicity properties were compared and analyzed in Table 3 and Table 4. The fifteen screened compounds have molecular weights less than 500, which suggest that they are likely to be absorbed and are able to reach the target when administered as drugs [56] and also their cLogP values lower than 5 that suggests a good absorption and permeation across cell membranes [56]. The drug score, cLogP, drug likeness and molar refractivity of substrate GMP (Id: DB01972) are 0.46, -5.97, -21.05, and 76.41 respectively. Interestingly, the compound DB03315, DB03593, DB04457, ChEMBL752, OMG, ZINC_3869846 and CHEMBL460901 have drug score value within 0.46 to 0.47 that is very close to substrate GMP, but high drug score (compare to drug score of GMP) value with 0.82 are observed in molecule ZINC_2522481 and ZINC_4963099. Subsequently, 1MG, CHEMBL582887, ZINC_13531931 and ZINC_4228242 compounds have lower drug score value (0.27 to 0.45) as compare to GMP. Moreover, predicted toxicity properties like mutagenic, tumorigenic, irritant and reproductive effective toxicity risks for all ligands have low value except ZINC_13531931 which was predicted to have a high reproductive effective toxicity risk. The highest value (4.43) of synthetic accessibility was observed for ligand 1MG, suggesting that it will be the slightly more difficult to synthesize from the compound library. Generally, the synthetic accessibility score (4.01 to 4.43) of all the ligands was within the range of moderate synthetic accessibility. It is also interesting to note that GMP, DB03315, DB03593 and GAO compounds not follow the Lipinski rule of five. Hence the remaining compounds may be considered as good lead compounds for isoform specific drug design.

3.6. Molecular Docking analysis

The identification of structural analogs of GMP or IMP may act as a good inhibitor that will reduce the molecular mechanisms and biochemical activity of substrate and product binding pocket of the enzyme and therefore they may lead to discovery of new drug targets. After critically assessment the pharmacological and toxicological properties of fifteen screened compounds, we have followed the following parameters for further molecular docking study (i) the compounds should follow Lipinski rule of five; (ii) drug score $>$ or $= 0.46$; and (iii) synthetic accessibility score < 4.43. Using the above mentioned criteria only eight compounds (DB04457, CHEMBL752, CHEMBL460901, OMG, ZINC_2522481, ZINC_4963099 and ZINC_3869846) from fifteen ligands were considered for molecular docking study. A comparative docking analysis was carried on the vicinity of binding affinity of hGMPR-I (PDB Id: 2BLE) and II (PDB Id: 2C6Q) crystal structures.

The docking result of those eight ligands and their hydrogen bonds interactions are mentioned in Table 5. In hGMPR-I, these ligands make contacts with Gly183, Ser184, Val185, Cys186, Asp219, Gly220, Cys222, Met240, Gly242, Gly243, Met269, Ser270 and Glu289 residues and their binding energies are within the range from -7.10 to -8.70 kcal/mol but they bind Gly183, Ser184, Cys186, Asp219, Gly220, Gly221, Cys222, Gly242 and Gly243 residue in hGMPR-II and their binding energies are within -7.90 to -9.30 to kcal/mol. The DB04457, CHEMBL752, CHEMBL460901, OMG, ZINC_3869846 molecules have drug score ~ 0.46 and their binding energies are lower in hGMPR-II compare to I. Interestingly, the molecule ZINC_2522481 and ZINC_4963099 have better drug score (0.82) compare to drug score of substrate GMP (0.46) and their binding energies are also reasonably better in isoform-II (-8.10 kcal/mol) rather than isoform-I (7.10/-7.60 kcal/mol). These docking results have demonstrated that these two compounds bind tightly in hGMPR-II near substrate binding site whereas they loosely bind in hGMPR-I at GMP binding site. The docked conformation of ZINC_2522481 and ZINC_4963099, with both isoforms, in the active sites of GMP is presented in Figure 10 and 11. Therefore, these two ligands may possible be considered as a better drug candidate than reaming six compounds whose drug score are within 0.46 that is very close to substrate GMP. Finally our theoretical results are based on critical observation on structural data analysis and are implemented in this work with valid justification. Therefore the present computational approach will definitely provide the light for developing isoform specific drug for the treatment of CML cancer and may be synthesized for future experimental and clinical studies.

27

Table 3: Pharmacokinetics properties of structural analogs of GMP/IMP using OSIRIS Property Explorer and Swiss ADME

Sl.no	LIGAND Id	Molecular Formula	OSIRIS Property Explorer							Swiss ADME		
			cLogP	Solubility	Molweight (gm/mol)	TPSA	Drug likeness	Drug-score	Molar Refractivity	Lipinski	Bioavailability Score	Synthetic Accessibility
1	DB01972 (GMP)	$C_{10}H_{14}N_5O_8P$	-5.97	-0.14	363	211.5	-21.05	0.46	76.41	No	0.11	4.28
2	DB03315	$C_{10}H_{14}N_5O_8P$	-5.97	-0.14	363	211.5	-19.73	0.46	76.41	No	0.11	4.28
3	DB03593	$C_{10}H_{14}N_5O_8P$	-5.97	-0.14	363	211.5	-19.73	0.46	76.14	No	0.11	4.28
4	DB04457	$C_{10}H_{14}N_5O_7P$	-5.43	-0.88	347	191.3	-16.57	0.46	75.25	Yes	0.11	4.07
5	CHEMBL752	$C_9H_{13}N_2O_9P$	-5.15	-0.74	347	195.8	-26.30	0.46	73.58	Yes	0.11	4.35
6	CHEMBL460901	$C_9H_{13}N_2O_9P$	-5.79	0.99	324	175.6	-20.32	0.47	65.18	Yes	0.11	4.25
7	CHEMBL582887	$C_9H_{14}N_5O_8P$	4.07	-5.19	416	63.21	-9.50	0.27	120.39	Yes	0.55	4.36
8	GAO	$C_{10}H_{14}N_5O_8P$	-5.97	0.14	363	211.5	-21.00	0.46	76.41	No	0.11	4.30
9	1MG	$C_{11}H_{16}N_5O_8P$	-5.72	0.22	377	202.7	-17.04	0.45	81.31	Yes	0.11	4.43
10	OMG	$C_{10}H_{14}N_5O_7P$	-5.54	0.27	377	200.5	-17.00	0.46	75.25	Yes	0.11	4.09
11	ZINC_2522481	$C_{11}H_{15}ClN_5O_5$	-2.26	-1.85	297	144.2	1.09	0.82	70.23	Yes	0.55	4.01
12	ZINC_4963099	$C_{11}H_{15}N_5O_5$	-2.26	-1.85	297	144.2	1.09	0.82	70.23	Yes	0.55	4.01
13	ZINC_13531931	$C_{10}H_{13}N_4O_8P$	-5.16	-0.37	348	190.0	-23.32	0.37	71.20	Yes	0.11	4.27
14	ZINC_3869846	$C_{10}H_{14}N_5O_7P$	-5.12	-0.54	347	191.3	-19.37	0.46	75.25	Yes	0.11	4.07
15	ZINC_4228242	$C_{10}H_{13}N_4O_8P$	-5.68	-0.04	348	185.5	-18.20	0.37	72.01	Yes	0.11	4.25

Table 3: Pharmacokinetics properties of structural analogs of GMP/IMP using OSIRIS Property Explorer and Swiss ADME

Abbreviation: ADME, absorption, distribution, metabolism excretion; TPSA, total polar surface area.

28

Sl.No.	LIGAND Id	Toxicity Risks			
		Mutagenic	Tumorigenic	Irritant	Reproductive effective
1	**DB01972 (GMP)**	Low	Low	Low	Low
2	DB03315	Low	Low	Low	Low
3	DB03593	Low	Low	Low	Low
4	DB04457	Low	Low	Low	Low
5	CHEMBL752	Low	Low	Low	Low
6	CHEMBL460901	Low	Low	Low	Low
7	CHEMBL582887	Low	Low	Low	Low
8	GAO	Low	Low	Low	Low
9	1MG	Low	Low	Low	Low
10	OMG	Low	Low	Low	Low
11	ZINC_2522481	Low	Low	Low	Low
12	ZINC_4963099	Low	Low	Low	Low
13	ZINC_13531931	Low	Low	Low	**High**
14	ZINC_3869846	Low	Low	Low	Low
15	ZINC_4228242	Low	Low	Low	Low

Table 4: Toxicity risks of structural analogs of GMP/IMP compounds predicted using OSIRIS Property Explorer

Sl.no.	Ligand Id	H-bond residue mediated Interactions		Binding Energy (kcal/mol)	
	(GMP/IMP)	hGMPR-I	hGMPR-II	hGMPR-I	hGMPR-II
1	DB04457	G183, S184,C186, D219, G220, C222, G242, G243, M269, E289	N158, S184, C186, T188, D219, G221, C222, G242, G243, E289	-8.20	-9.10
2	CHEMBL752	G183, S184, V185, C186, G242, G243, S270	S184, C186, T188, D219, G221, C222, G242, G243	-8.10	-8.80
3	CHEMBL460901	C186, T188, D219, G242, M269, S270, E289	G183, S184, V185, C186, D219, C222, M240, G242, G243, M269	-7.60	-7.90
5	OMG	A131, G179, G181, G183, S270, M269, G289	A131, N158, G179, G183, S184, T188, D219, G220, G221, L241, G242, G243	-7.40	-8.60
6	ZINC_2522481	G183, S184, V185, C186, D219, M240, G268, M269, S270, E289, G290	D129, N158, K177, G183, S184, V185, C186, D219, G220, G242, G243	-7.10	-8.10
7	ZINC_4963099	N158, G179, V180, S184, V185, C186, D219, M269, E289	D129, N158, K177, G179, G183, S184, V185, C186, D219, G220, G242	-7.60	-8.10
8	ZINC_3869846	G181, V180, G179, N158, G183, S184, V185, C186, D219, G220, M240, L241, G242, G243	N158, G179, G181, D219, G220, G221, C222, M240, G242, G243, M244	-8.50	-9.30

Table 5: Molecular docking analysis of structural analogs of GMP/IMP in hGMPR-I and II

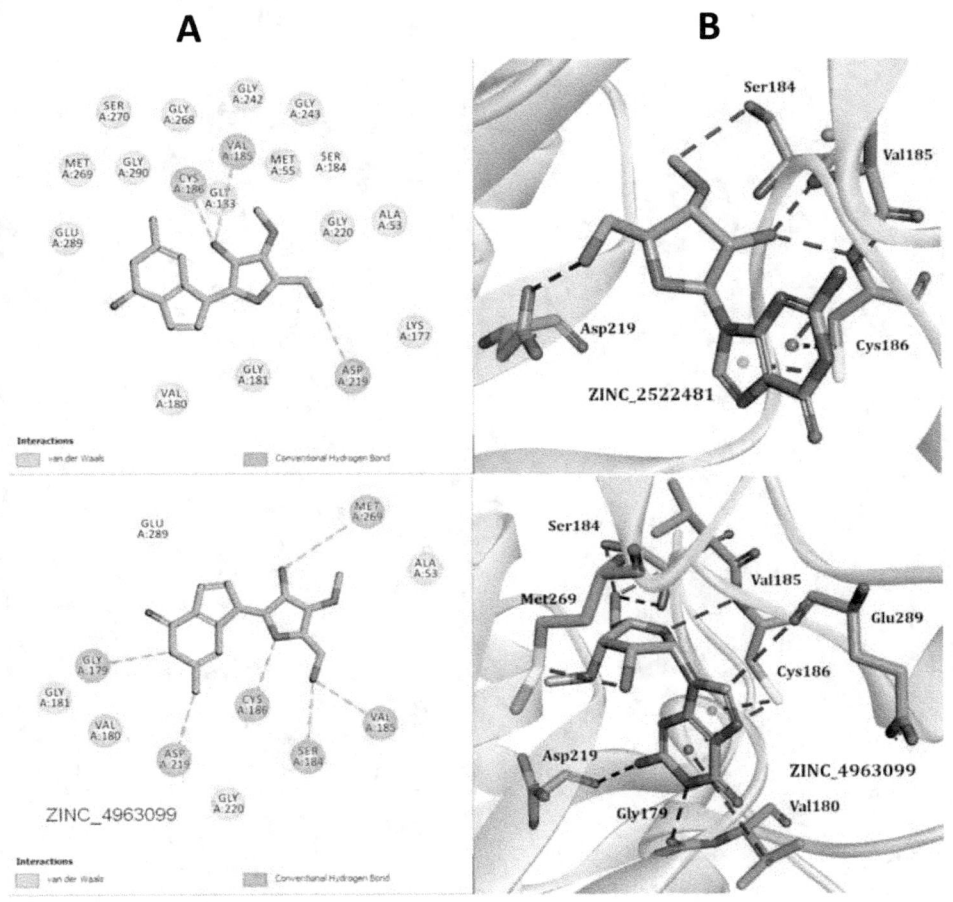

Figure 10: Molecular docking interactions between ZINC_2522481 and ZINC_4963099 of hGMPR-I (PDB Id: 2BLE): (A) 2-D model of the ligand-residue interactions; (B) 3-D model of the ligand-residue interactions

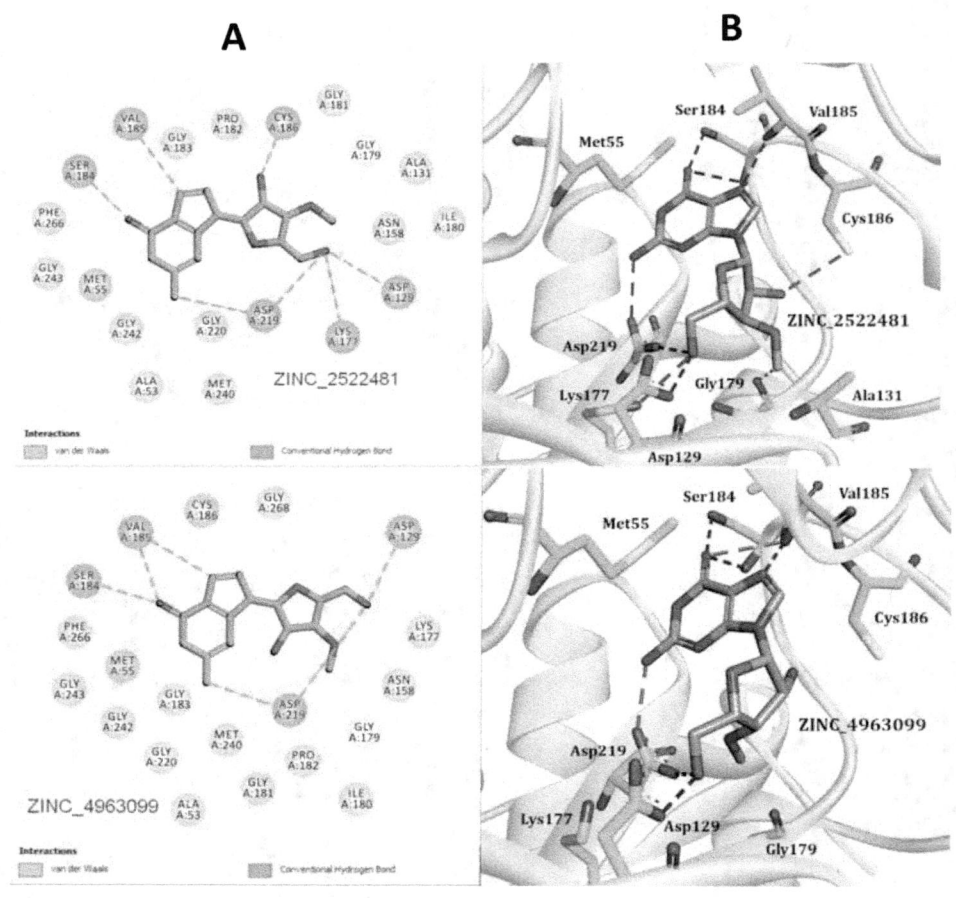

Figure 11: Molecular docking interactions between ZINC_2522481 and ZINC_4963099 of hGMPR-II (PDB Id: 2C6Q): (A) 2-D model of the ligand-residue interactions; (B) 3-D model of the ligand-residue interactions

4. CONCLUSION

The MD-results suggest the position of GMP binding site which is occupied by four conserved water molecules in unliganded conformation of hGMPR enzyme (Figure 12) but no crystal structure is available to represent the native state. Interestingly, these water molecules also stabilize that pocket by Cys186---W1---W2---W3---W4---Glu289/Thr188 interaction. The W1, W2, W3 and W4 hydrophilic positions are generally occupied either by substrate (GMP) or by product (IMP) in ligand bound state. Possibly, four conserved water molecules may inactivate GMP binding pocket in an unliganded state but they are departed from those concerning positions when enzyme (hGMPR)-substrate complex is formed. So, MD results have confirmed the significant role of four conserved water molecules at catalytic region of GMP binding site (Figure 13) and their active participation in recognition and stabilization of substrate binding pocket with catalytic residues which may be inferred towards biochemical insights of hGMPR enzyme. The stereochemical features and topologies of these conserved water molecules (W1, W2, W3 and W4) and their positions in the catalytic pocket, may also facilitate future drug discovery by design of appropriately oriented chemical groups to mimic the structural and electronic properties of conserved water molecules. Hence, virtual screening technique and molecular docking study has been applied on the crystal structures of hGMPR enzyme (as a receptor) to investigate the structural analog of substrate GMP, and the results suggest compound ZINC_2522481 and ZINC_4963099 have a better drug score compare to substrate GMP. Both these compounds bind tightly at substrate bind sites in human GMPR-II than I, and they may act as a better isoform specific drug candidate for CML cancer protein.

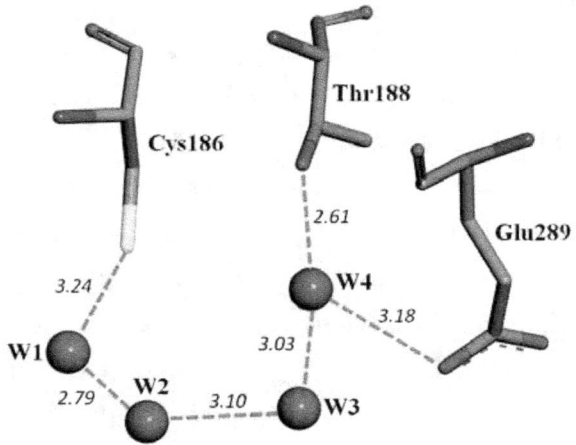

Figure 12: The water mediated H- bonding interaction of Cys186 with Thr188 and Glu289 in native conformation MD structure of hGMPR enzyme

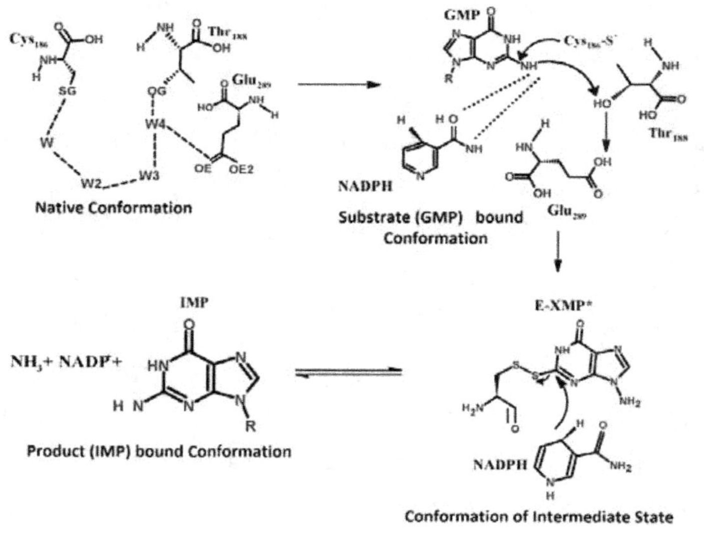

Figure 13: Conformational transition in GMP binding domain of hGMPR enzyme from native to product bound conformation

ACKNOWLEDGMENT: H.R.B is thankful to Council of Scientific & Industrial Research (CSIR) for providing financial support of Senior Research Associateship (Pool Scientist scheme) of the Government of India. We thank Supercomputing Facility for Bioinformatics & Computational Biology at Indian Institute of Technology-New Delhi and Department of Computer Science at Jamia Millia Islamia for providing the computing resources.

AUTHOR'S CONTRIBUTION: H.R.B. supervised all computational experiment, conceived and designed the study and drafted the entire manuscript. The authors G.P.R, A.T. and S.R. contributed equally to this manuscript. They had performed the MD simulation, literature search, data analysis and reviewed manuscript.

LITERATURE:

1. Hehlmann, R., Hochhaus, A. and Baccarani, M. (2007). *Chronic myeloid leukaemia.* The Lancet, 370(9584), pp.342-350.

2. Ganesan, P. and Kumar, L. (2017). *Chronic myeloid leukemia in India.* Journal of global oncology, 3(1), pp.64-71.

3. Cortez, D., Stoica, G., Pierce, J.H. and Pendergast, A.M. (1996). *The BCR-ABL tyrosine kinase inhibits apoptosis by activating a Ras-dependent signaling pathway.* Oncogene, 13(12), pp.2589-2594.

4. Notari, M., Neviani, P., Santhanam, R., Blaser, B.W., Chang, J.S., Galietta, A., Willis, A.E., Roy, D.C., Caligiuri, M.A., Marcucci, G. and Perrotti, D. (2006). *A MAPK/HNRPK pathway controls BCR/ABL oncogenic potential by regulating MYC mRNA translation.* Blood, 107(6), pp.2507-2516.

5. Neshat, M.S., Raitano, A.B., Wang, H.G., Reed, J.C. and Sawyers, C.L. (2000). *The survival function of the Bcr-Abl oncogene is mediated by Bad-dependent and-independent pathways: roles for phosphatidylinositol 3-kinase and Raf.* Molecular and cellular biology, 20(4), pp.1179-1186.

6. Sawyers, C.L. (1993). *The role of myc in transformation by BCR-ABL.* Leukemia & lymphoma, 11(sup1), pp.45-46.

7. Steelman, L.S., Pohnert, S.C., Shelton, J.G., Franklin, R.A., Bertrand, F.E. and McCubrey, J.A. (2004). *JAK/STAT, Raf/MEK/ERK, PI3K/Akt and BCR-ABL in cell cycle progression and leukemogenesis.* Leukemia, 18(2), pp.189-218.

8. Jabbour, E. and Kantarjian, H. (2018). *Chronic myeloid leukemia: 2018 update on diagnosis, therapy and monitoring.* American journal of hematology, 93(3), pp.442-459.

9. World Health Organization. (2014). *Chronic Myelogenous Leukemia.* Retrieved from:https://www.who.int/selection_medicines/committees/expert/20/applicatio ns/CML.pdf?ua=1

10. Bairagya, H.R. and Bansal, M. (2016). *New insight into the architecture of oxy-anion pocket in unliganded conformation of GAT domains: A MD-simulation study.* Proteins: Structure, Function, and Bioinformatics, 84(3), pp.360-373.

11. Zhang, J., Zhang, W., Zou, D., Chen, G., Wan, T., Zhang, M. and Cao, X. (2003). *Cloning and functional characterization of GMPR2, a novel human guanosine monophosphate reductase, which promotes the monocytic differentiation of HL-60 leukemia cells.* Journal of cancer research and clinical oncology, 129(2), pp.76-83.

12. Hedstrom, L. (2012). *The dynamic determinants of reaction specificity in the IMPDH/GMPR family of (β/α) 8 barrel enzymes.* Critical reviews in biochemistry and molecular biology, 47(3), pp.250-263.

13. Patton, G.C., Stenmark, P., Gollapalli, D.R., Sevastik, R., Kursula, P., Flodin, S., Schuler, H., Swales, C.T., Eklund, H., Himo, F. and Nordlund, P. (2011). *Cofactor mobility determines reaction outcome in the IMPDH and GMPR (β-α) 8 barrel enzymes.* Nature chemical biology, 7(12), p.950.

14. Hedstrom, L. (2009). *IMP dehydrogenase: structure, mechanism, and inhibition.* Chemical reviews, 109(7), pp.2903-2928.

15. Knight, J.D., Hamelberg, D., McCammon, J.A. and Kothary, R. (2009). *The role of conserved water molecules in the catalytic domain of protein*

kinases. Proteins: Structure, Function, and Bioinformatics, 76(3), pp.527-535.

16. de Beer, S., Vermeulen, N.P. and Oostenbrink, C. (2010). *The role of water molecules in computational drug design.* Current topics in medicinal chemistry, 10(1), pp.55-66.

17. Bairagya, H.R., Mukhopadhyay, B.P. and Sekar, K. (2009). *Conserved water mediated H-bonding dynamics of inhibitor, cofactor, Asp 364 and Asn 303 in human IMPDH II.* Journal of Biomolecular Structure and Dynamics, 26(4), pp.497-507.

18. Bairagya, H.R., Mukhopadhyay, B.P. and Sekar, K. (2009). *An insight to the dynamics of conserved water molecular triad in IMPDH II (human): Recognition of cofactor and substrate to catalytic Arg 322.* Journal of Biomolecular Structure and Dynamics, 27(2), pp.149-158.

19. Bairagya, H.R., Mukhopadhyay, B.P. and Bera, A.K. (2011). *Conserved water mediated recognition and the dynamics of active site Cys 331 and Tyr 411 in hydrated structure of human IMPDH-II.* Journal of Molecular Recognition, 24(1), pp.35-44.

20. Bairagya, H.R., Mukhopadhyay, B.P. and Bera, A.K. (2011). *Role of salt bridge dynamics in inter domain recognition of human IMPDH isoforms: an insight to inhibitor topology for isoform-II.* Journal of Biomolecular Structure and Dynamics, 29(3), pp.441-462.

21. Karplus, M. and Petsko, G.A. (1990). *Molecular dynamics simulations in biology.* Nature, 347(6294), pp.631-639.

22. Bernstein, F.C., Koetzle, T.F., Williams, G.J., Meyer Jr, E.F., Brice, M.D., Rodgers, J.R., Kennard, O., Shimanouchi, T. and Tasumi, M. (1977). *The Protein Data Bank: A computer-based archival file for macromolecular structures.* European journal of biochemistry, 80(2), pp.319-324.

23. Guex, N., Diemand, A., Peitsch, M.C. and Schwede, T. (2000). *SwissPDBViewer program.* Glaxo Smith Kline R&D, Brentford.

24. Apweiler, R., Bairoch, A., Wu, C.H., Barker, W.C., Boeckmann, B., Ferro, S.,

Gasteiger, E., Huang, H., Lopez, R., Magrane, M. and Martin, M.J. (2004). *UniProt: the universal protein knowledgebase.* Nucleic acids research, 32(suppl_1), pp.D115-D119.

25. Sievers, F. and Higgins, D.G. (2014). *Clustal omega.* Current protocols in bioinformatics, 48(1), pp.3-13.

26. Booth, A.D. (1947). *Application of the method of steepest descents to X-ray structure analysis.* Nature, 160(4058), pp.196-196.

27. Daniel, J.W. (1967). *Convergence of the conjugate gradient method with computationally convenient modifications.* Numerische Mathematik, 10(2), pp.125-131.

28. Pisabarro, M.T., Ortiz, A.R., Serrano, L. and Wade, R.C. (1994). *Homology modeling of the Abl-SH3 domain.* Proteins: Structure, Function, and Bioinformatics, 20(3), pp.203-215.

29. Humphrey, W., Dalke, A. and Schulten, K. (1996). *VMD: visual molecular dynamics.* Journal of molecular graphics, 14(1), pp.33-38.

30. Cerutti, D.S., Jain, T. and McCammon, J.A. (2006). *CIRSE: A solvation energy estimator compatible with flexible protein docking and design applications.* Protein science, 15(7), pp.1579-1596.

31. Dey, P., Bairagya, H.R. and Roy, A. (2018). *Putative role of invariant water molecules in the X-ray structures of family G fungal endoxylanases.* Journal of biosciences, 43(2), pp.339-349.

32. Huang, J. and MacKerell Jr, A.D. (2013). *CHARMM36 all-atom additive protein force field: Validation based on comparison to NMR data.* Journal of computational chemistry, 34(25), pp.2135-2145.

33. Manoharan, P. and Ghoshal, N. (2018). *Fragment-based virtual screening approach and molecular dynamics simulation studies for identification of BACE1 inhibitor leads.* Journal of Biomolecular Structure and Dynamics, 36(7), pp.1878-1892.

34. Kalé, L., Skeel, R., Bhandarkar, M., Brunner, R., Gursoy, A., Krawetz, N.,

Phillips, J., Shinozaki, A., Varadarajan, K. and Schulten, K. (1999). *NAMD2: greater scalability for parallel molecular dynamics*. Journal of Computational Physics, 151(1), pp.283-312.

35. Sumathi, K., Ananthalakshmi, P., Roshan, M.M. and Sekar, K. (2006). *3dSS: 3D structural superposition*. Nucleic Acids Research, 34(suppl_2), pp.W128-W132.

36. Dey, P., Bairagya, H.R. and Roy, A. (2018). *Putative role of invariant water molecules in the X-ray structures of family G fungal endoxylanases*. Journal of biosciences, 43(2), pp.339-349.

37. Wishart, D.S., Feunang, Y.D., Guo, A.C., Lo, E.J., Marcu, A., Grant, J.R., Sajed, T., Johnson, D., Li, C., Sayeeda, Z. and Assempour, N. (2018). *DrugBank 5.0: a major update to the DrugBank database for 2018*. Nucleic acids research, 46(D1), pp.D1074-D1082.

38. Zoete, V., Daina, A., Bovigny, C. and Michielin, O. (2016). *SwissSimilarity: a web tool for low to ultra high throughput ligand-based virtual screening.*

39. Feng, Z., Chen, L., Maddula, H., Akcan, O., Oughtred, R., Berman, H.M. and Westbrook, J. (2004). *Ligand Depot: a data warehouse for ligands bound to macromolecules*. Bioinformatics, 20(13), pp.2153-2155.

40. Davies, M., Nowotka, M., Papadatos, G., Dedman, N., Gaulton, A., Atkinson, F., Bellis, L. and Overington, J.P. (2015). *ChEMBL web services: streamlining access to drug discovery data and utilities*. Nucleic acids research, 43(W1), pp.W612-W620.

41. Sterling, T. and Irwin, J.J. (2015). *ZINC 15–ligand discovery for everyone*. Journal of chemical information and modeling, 55(11), pp.2324-2337.

42. Lipinski, C.A., Lombardo, F., Dominy, B.W. and Feeney, P.J. (1997). *Experimental and computational approaches to estimate solubility and permeability in drug discovery and development settings*. Advanced drug delivery reviews, 23(1-3), pp.3-25.

43. Behrouz, S., Rad, M.N.S., Shahraki, B.T., Fathalipour, M., Behrouz, M. and Mirkhani, H. (2019). *Design, synthesis, and in silico studies of novel eugenyloxy*

propanol azole derivatives having potent antinociceptive activity and evaluation of their β-adrenoceptor blocking property. Molecular diversity, 23(1), pp.147-164.

44. Sander, T. (2001). *OSIRIS property explorer.* Organic Chemistry Portal. Actelion Pharmaceuticals Ltd.: Allschwil, Switzerland. Retrieved from http://www.organic-chemistry.org/prog/peo/.

45. Daina, A., Michielin, O. and Zoete, V. (2017). *SwissADME: a free web tool to evaluate pharmacokinetics, drug-likeness and medicinal chemistry friendliness of small molecules.* Scientific reports, 7, p.42717.

46. Dar, A.M. and Mir, S. (2017). *Molecular docking: approaches, types, applications and basic challenges.* J Anal Bioanal Tech, 8(2), pp.1-3.

47. Kitchen, D.B., Decornez, H., Furr, J.R. and Bajorath, J. (2004). *Docking and scoring in virtual screening for drug discovery: methods and applications.* Nature reviews Drug discovery, 3(11), pp.935-949.

48. Ewing, T.J., Makino, S., Skillman, A.G. and Kuntz, I.D. (2001). *DOCK 4.0: search strategies for automated molecular docking of flexible molecule databases.* Journal of computer-aided molecular design, 15(5), pp.411-428.

49. Wang, R., Lai, L. and Wang, S. (2002). *Further development and validation of empirical scoring functions for structure-based binding affinity prediction.* Journal of computer-aided molecular design, 16(1), pp.11-26.

50. Gohlke, H., Hendlich, M. and Klebe, G. (2000). *Knowledge-based scoring function to predict protein-ligand interactions.* Journal of molecular biology, 295(2), pp.337-356.

51. Trott, O. and Olson, A.J. (2010). *AutoDock Vina: improving the speed and accuracy of docking with a new scoring function, efficient optimization, and multithreading.* Journal of computational chemistry, 31(2), pp.455-461.

52. Morris, G.M., Huey, R., Lindstrom, W., Sanner, M.F., Belew, R.K., Goodsell, D.S. and Olson, A.J. (2009). *AutoDock4 and AutoDockTools4: Automated docking with selective receptor flexibility.* Journal of computational

chemistry, 30(16), pp.2785-2791.

53. Morris, G.M. and Lim-Wilby, M. (2008). *Molecular docking.* In Molecular modeling of proteins (pp. 365-382). Humana Press.

54. Goodsell, D.S., Morris, G.M. and Olson, A.J. (1996). *Automated docking of flexible ligands: applications of AutoDock.* Journal of molecular recognition, 9(1), pp.1-5.

55. Wu, C.Y. and Benet, L.Z. (2005). *Predicting drug disposition via application of BCS: transport/absorption/elimination interplay and development of a biopharmaceutics drug disposition classification system.* Pharmaceutical research, 22(1), pp.11-23.

TABLE OF CONTENTS

Publisher: Eliva Press SRL

Email: info@elivapress.com

All rights reserved.

Eliva Press is an independent publishing house established for the publication and dissemination of academic works all over the world. Company provides high quality and professional service for all of our authors.

Our Services:
Free of charge, open-minded, eco-friendly, innovational.

-All services are free of charge for you as our author (manuscript review, step-by-step book preparation, publication, distribution, and marketing).
-No financial risk. The author is not obliged to pay any hidden fees for publication.
-Editors. Dedicated editors will assist step by step through the projects.
-Money paid to the author for every book sold. Up to 50% royalties guaranteed.
-ISBN (International Standard Book Number). We assign a unique ISBN to every Eliva Press book.
-Digital archive storage. Books will be available online for a long time. We don't need to have a stock of our titles. No unsold copies. Eliva Press uses environment friendly print on demand technology that limits the needs of publishing business. We care about environment and share these principles with our customers.
-Cover design. Cover art is designed by a professional designer.
-Worldwide distribution. We continue expanding our distribution channels to make sure that all readers have access to our books.

www.elivapress.com

www.ingramcontent.com/pod-product-compliance
Lightning Source LLC
Chambersburg PA
CBHW071244220526
45468CB00002B/991